Mayfly

my life cycle

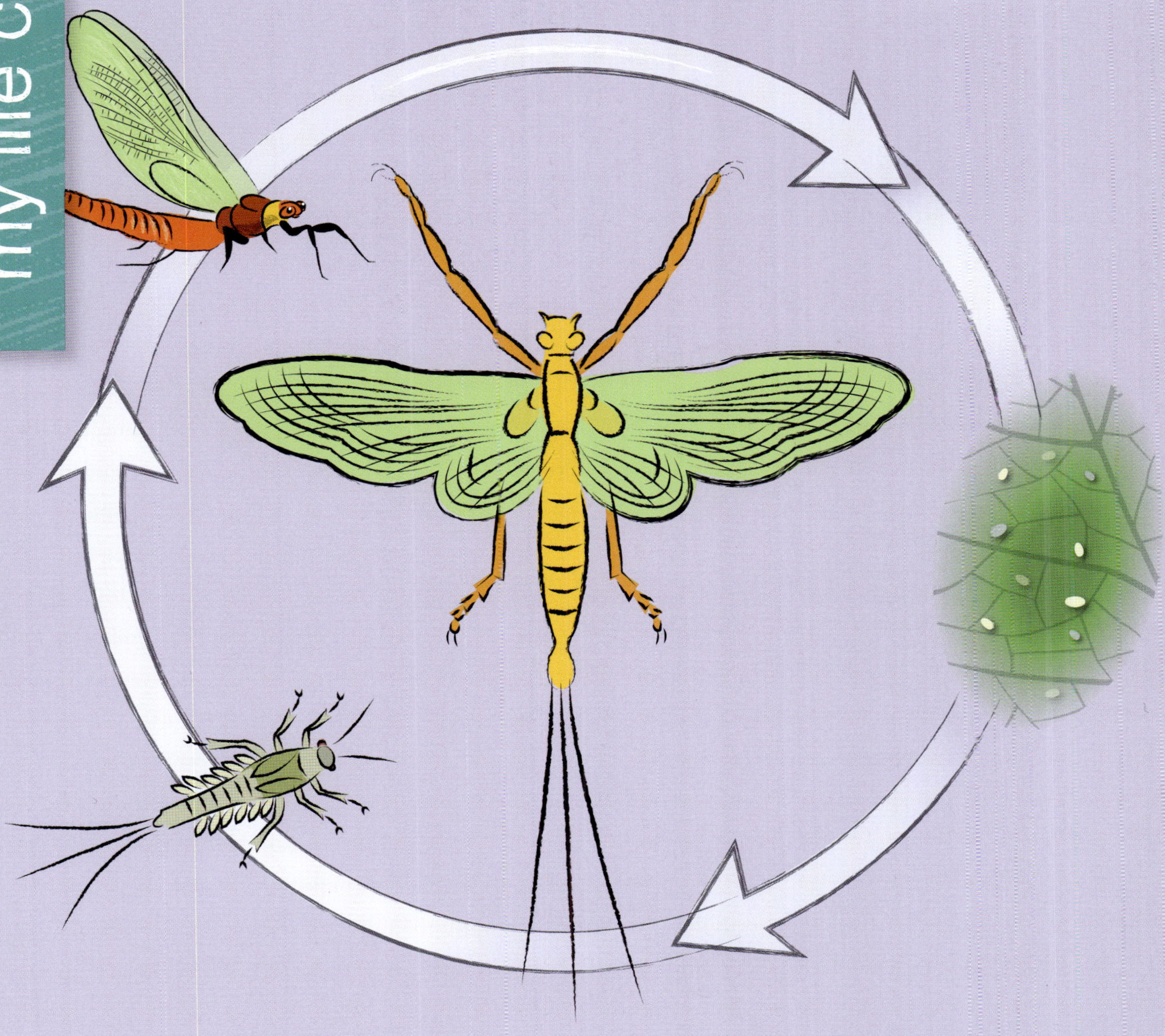

CHERRY LAKE PRESS

Published in the United States of America by Cherry Lake Publishing Group
Ann Arbor, Michigan
www.cherrylakepublishing.com

Reading Adviser: Marla Conn, MS, Ed., Literacy specialist, Read-Ability, Inc.
Book Designer: Jennifer Wahi
Illustrator: Jeff Bane

Photo Credits: © Bildagentur Zoonar GmbH/Shutterstock.com, 5; © Rostislav Stefanek/Shutterstock.com, 7, 9; © Bryan Neuswanger/Shutterstock.com, 11; © boyphare/Shutterstock.com, 13; © Sandra Standbridge/Shutterstock.com, 15; © Mladen Mitrinovic/Shutterstock.com, 17, 21; © Laky_981/Shutterstock.com, 19; © David Jeffery Morgan/Shutterstock.com, 23; Cover, 2-3, 6, 10, 22, 24, Jeff Bane

Cherry Lake Press is an imprint of Cherry Lake Publishing Group.

Library of Congress Cataloging-in-Publication Data

Names: Gray, Susan Heinrichs, author. | Bane, Jeff, 1957- illustrator.
Title: Mayfly / Susan H. Gray ; illustrated by Jeff Bane.
Description: Ann Arbor, Michigan : Cherry Lake Publishing, 2021. | Series: My life cycle | Includes index. | Audience: Grades 2-3 (provided by Cherry Lake Publishing)
Identifiers: LCCN 2020030678 (print) | LCCN 2020030677 (ebook) | ISBN 9781534180031 (hardcover) | ISBN 9781534181748 (paperback) | ISBN 9781534181045 (pdf) | ISBN 9781534182752 (ebook)
Subjects: LCSH: Mayflies--Life cycles--Juvenile literature.
Classification: LCC QL505 .G73 2021 (ebook) | LCC QL505 (print) | DDC 595.7/34156--dc23
LC record available at https://lccn.loc.gov/2020030678

Printed in the United States of America
Corporate Graphics

table of contents

About the author: Susan H. Gray has a master's degree in zoology. She loves writing science books, especially about animals. Susan lives in Arkansas with her husband, Michael. They have seen mayflies in Arkansas for many years.

About the illustrator: Jeff Bane and his two business partners own a studio along the American River in Folsom, California, home of the 1849 Gold Rush. When Jeff's not sketching or illustrating for clients, he's either swimming or kayaking in the river to relax.

I am a mayfly egg resting in a stream. I'm so small, you can barely see me.

After I **hatch**, I'm a tiny **nymph**. I have six legs. I crawl under a rock to eat some **algae**.

To grow, I have to **molt**. First, my **exoskeleton** splits. Then, a bigger me slides out. In my life, I might molt 50 times!

One day, I climb out of the water to molt. This time, I come out with wings!

I’m a **subimago** now. I fly to a tree and rest a while.

After one last molt, I’m an **imago**. I’m a grown-up now.

A mayfly **swarm** is nearby, and I join in. There are millions of us!

I find a female, and we **mate**. Right away, she begins laying eggs.

Over and over, she dives toward the water. She dips her **abdomen** in, dropping eggs each time.

Someday, they will hatch into tiny nymphs.

glossary & index

glossary

abdomen (AB-duh-muhn) the rear section of an insect's body

algae (AL-jee) simple plants that live in ponds, streams, and other bodies of water

exoskeleton (ek-soh-SKEL-uh-tuhn) the hard outer covering that insects have

hatch (HACH) to break out of an egg

imago (ih-MAY-goh) the adult stage of an insect

mate (MATE) to join together to produce babies

molt (MOHLT) to shed an outer body covering in order to grow

nymph (NIMF) a young mayfly that is undergoing changes

subimago (sub-ih-MAY-goh) the stage between nymph and imago when a mayfly is able to fly

swarm (SWORM) a group of insects that gather or move in large numbers

index